Church, It's Time to Try AI:
Because the Marthas Want to Take a Seat Too

Church, It's Time to Try AI: Because the Marthas Want to Take a Seat Too
A Beginners Guide to Help Pastors and Ministry Leaders Stretch Time, Brainstorm Ideas and Leverage Resources

Copyright © 2025 MeShayle Lester
MeShelfies Books & Print
Lancaster, TX

All rights reserved. No part of this book may be reproduced, stored in a retrieval system, or transmitted in any form or by any means – electronic, mechanical, photocopy, recording, or otherwise – without prior written permission of the copyright owner.

Disclaimer Notice: Please note the information contained withing this document is for educational purposes only. All effort has been executed to present accurate, up to date, reliable, complete information. No warranties of any kind are declared or implied. Readers acknowledge that the author is not engaged in the rendering of legal, financial, medical or professional advice. The content within this book has been derived from various sources including AI. Please consult a licensed professional before attempting any technique outlines in this book. By reading this document, the reader agrees that under no circumstances is the author responsible for any losses, direct or indirect, that are incurred as a result of the use of the information contained within this document, including, but not limited to, errors, omissions, or inaccuracies.

Scripture quotations marked ESV are taken from ESV® Bible (The Holy Bible, English Standard Version®), copyright © 2001 by Crossway Bibles, a publishing ministry of Good News Publishers. Used by permission. All rights reserved.

Scripture quotations taken from The Holy Bible, New International Version®, NIV®. Copyright © 1973, 1978, 1984, 2011 by Biblica, Inc. Used with permission of Zondervan. All rights reserved worldwide. www.zondervan.com

Scripture quotations marked NLT are taken from the Holy Bible, New Living Translation, copyright © 1996, 2004, 2015 by Tyndale House Foundation. Used by permission of Tyndale House Publishers, Carol Stream, Illinois 60188. All rights reserved.

This book was created with the following AI tools:
ChatGPT 4o, ChatGPT 5

Cover Design and Graphics by MeShayle Lester using Flux 1.0 + Photoshop
Creative Direction by MeShayle Lester

Printed in the United States of America
ISBN: 978-1-7367154-6-8

1 2 3 4 5 6 7 8 9 10 - 0

D. MESHAYLE LESTER

Church, It's Time to Try AI:
Because the Marthas Want to Take a Seat Too

A Beginners Guide to Help Pastors and Ministry Leaders
Stretch Time, Brainstorm Ideas and Leverage Resources

Meshelfies
BOOKS & PRINT

Dedication

For siblings: Allie Rae, Tessie, and Lee Jr.

When I think of the three of you, I am reminded of Martha, Mary, and Lazarus.
Martha knew how to move things forward and make things happen.
Mary carried the spirit of worship and drew close to what mattered most.
And Lazarus, all he had to do was be in the room; his testimony alone was a gift.

A threefold cord is not easily broken. (Ecclesiastes 4:12)

This book was compiled with the assistance of AI,

using ChatGPT models 4o and 5.

The cover design was created with Flux 1.0 and refined in Photoshop.

CONTENTS

PART I - FOUNDATIONS & VISION

Chapter 1 – Why This? Why Now?.. 25

    Standing at the Crossroads of Tomorrow 25

Chapter 2 – Seeing Clearly: What AI Is and What It Is Not 29

    Understanding ChatGPT With Wisdom and Caution 29

PART II - LEADING THE CHURCH WELL

Chapter 3 – Putting the House in Order 35

    Aligning Administration With the Heart of Ministry................... 35

Chapter 4 – Caring for the Flock 41

    Member Care and Communication in a Digital World............... 41

Chapter 5 – Teaching That Transforms.. 45

    Discipleship & Education.. 45

Chapter 6 – Shepherding the Next Generation 49

    AI in Youth and Children's Ministry.. 49

Chapter 7 – The Streets We Serve 53

    Community Engagement 53

Chapter 8 – A Church for All: Accessibility, Inclusion, and AI 61

    Language, Disability, and the Gospel Without Barriers 61

PART III – STEWARDSHIP OF TIME, TALENT & TREASURE

Chapter 9 – Redeeming the Time 67

    Administration & Efficiency.. 67

INNOVATION, CREATIVITY & EMPOWERMENT

Chapter 10 – Singing the Lord's Song in a Digital Land 73

    Photos, Videos, Music and Media.. 73

Chapter 11 – Stewardship, Grant Writing, and Economic Empowerment 79

    Funding 79

Chapter 12 – Marketplace Ministry 87

    When Work Funds the Witness 87

PART IV – BUILDING THE FUTURE CHURCH

Chapter 13 – Vision Builders: Church Planting & Innovation.......... 95

    Embracing New Ground with Faith and Strategy....................... 95

Chapter 14 – Wisdom in a Wired World 99

    Spiritual Discernment & Guardrails 99

Chapter 15 – Walking Justly in the Age of AI 103

    Ethics, Responsibility, and Ministry Integrity 103

Chapter 16 – Watchmen on the Wall... 107

    Security, Privacy, and Digital Trust 107

BRIDGING THE PAST AND FUTURE

Chapter 17 – The Cloud of Witnesses: Preserving Our Stories..... 113

Testimonies and Elders in a Digital Age 113

Chapter 18 – Beyond the Horizon 117

Proclaiming the Word in a Digitally Discerning Age 117

Chapter 19 – God's Gift: Man, Not Machine 121

WHEN NOT TO USE AI 121

From the Editor

> *"As Jesus and his disciples were on their way, he came to a village where a woman named Martha opened her home to him. She had a sister called Mary, who sat at the Lord's feet listening to what he said.*
>
> *But Martha was distracted by all the preparations that had to be made. She came to him and asked, "Lord, don't you care that my sister has left me to do the work by myself? Tell her to help me!"*
>
> *"Martha, Martha," the Lord answered, "you are worried and upset about many things, but few things are needed - or indeed only one. Mary has chosen what is better, and it will not be taken away from her."*
> *— Luke 10:38-42*

I have been walking alongside pastors since my teenage years. Long before the degrees and titles, I helped carry the vision, making sure things did not fall through the cracks, and learning ministry by holding it up from behind the scenes. Over the years, I have served as a surrogate leader, trusted to keep the work moving forward while others carried the pulpit.

By my late twenties, God called me deeper into prayer, into intercession, into a level of worship that showed me something I'll never forget: worship is greater than work. We serve a God who speaks, and when He made it plain to me, it was clear: He wanted my heart more than my hands.

But here is the tension: I was still living the life of "Martha". Always busy. Always needed. Always keeping things going. Like so many of our mothers and workers in the church, I knew what it meant to be faithful getting the work done, while longing for the chance to be seated in His presence. And let me just be honest: if the technology we have now had been available then? My Lord, I could have gotten so much more done with so much less strain. Tasks that wore me out could have been automated. Details that stole my prayer time could have been handled in minutes. The church should become literate in AI, because the "Marthas" want to take a seat too.

That is why this book matters. I write as a servant who knows both sides: the study and the struggle, the call to prayer and the pull of paperwork. I have built websites, designed flyers, managed finances, served as a youth and marketing director, developed lesson plans, facilitated leadership meetings, produced my pastor's radio show, and spent many late nights after service to make sure God's house ran right.

Rabbi Ben Avraham once asked, *"If you cannot hear [Jesus] when He calls, how can you answer?"* That word has never left me. The key is to always have an ear to hear. Never be so busy, so burdened, or so consumed with ministry tasks that you cannot hear the Master's voice. AI has the potential to distract us from God's voice, but it can also be harnessed as a tool to help us listen to Him more deeply.

This book is about redeeming time and equipping the saints. It is about reminding us that technology, when rightly harnessed, can serve Kingdom purposes. Not to replace the Spirit, but to release God's people into greater capacity for Kingdom work. Not to diminish worship, but to create more space for it.

So, to every "Martha" who feels stretched thin: I see you, because I *am* you. And my prayer is that the use of AI – when done right – helps you reclaim your time, reclaim your strength, and reclaim your seat in His presence. Because in the end, worship will always be greater than work.

Dr. D. MeShayle Lester

Introduction

A New Chapter in an Old Story

Let me begin with a simple truth: technology has always been part of the Church's story. From the earliest days, God's people have learned to use whatever tools were available to carry His message forward.

During the fifteenth century, the Bible was one of the first major works to be printed on a press. For the first time, the Word of God was no longer confined to handwritten manuscripts stored in monasteries. Ordinary believers could hold Scripture in their hands and read it for themselves. That technology, once viewed with suspicion by some, became a spark that ignited the Reformation and changed the course of history.

Centuries later, radio waves carried the gospel into homes and across nations. Missionaries and evangelists began to speak not just to crowds in tents or on street corners, but to millions of people gathered around radios in living rooms. Then came television, which brought worship services and preaching into households worldwide. More recently, livestreaming has allowed churches to stay connected across time zones, pandemics, and distance. These tools have reminded us that the gospel cannot be chained.

Every generation faces the same question: Will we fear new tools, or will we learn to steward them for God's glory?

Artificial intelligence is simply the next tool in that line. It may feel new, overwhelming, or even intimidating, but in the end, it is still just a tool. Like every tool before it, its value depends on the hands that hold it and the heart that guides it.

Why This Book Exists

This book is not about replacing pastors or spiritual leaders. It is not about outsourcing prayer, preaching, or shepherding to a machine. No algorithm can shepherd a soul, preach with Spirit-filled conviction, or love people the way Jesus has called us to love.

This book is about equipping *you*.

My desire is to show how artificial intelligence can come alongside your ministry to ease the burden, spark creativity, and open new doors for connection. Used wisely, these tools can free up your time so you can focus more deeply on what only you can do: prayer, presence, discipleship, vision, and leadership.

Think of AI not as a replacement, but as an amplifier. These tools can help multiply your efforts without diluting your calling. They are not here to take your place. They are here to support your mission.

What You Will Learn

In these chapters, you will discover:

- **How AI works**, explained in plain language. You do not need to be a tech expert. I will walk you through what models are, how prompts work, and why ethical concerns like bias matter.

- **Practical applications for ministry.** From administration and teaching, to outreach and member care, you will find real examples of how AI can assist without overstepping.

- **Spiritual guardrails.** We will speak honestly about the limitations, risks, and necessary boundaries for using these tools with integrity.

- **Reflection and action.** Many chapters end with workbook-style questions, prayers, or exercises to help you apply what you are learning to your own context.

- **Prompts you can use today.** Throughout the book, you will see sample prompts not just for ChatGPT, but for other tools as well. Even as technology evolves, the core principles of thoughtful use will remain the same.

While I am deeply excited about the possibilities of AI, I want to offer a clear and pastoral word of warning. No tool, no matter how advanced, can replace the Spirit of God.

AI cannot discern truth. It cannot bring conviction, nor can it guide you into wisdom. It cannot replace Scripture, prayer, or the quiet leading of the Holy Spirit.

As you explore these technologies, use them, but do not lean on them. Learn from them, but do not bow to them. Let them serve you, not lead you. Remember that your source of power is not in the machine, but in the presence of God.

A Beginners Guide

AI is constantly evolving, with new capabilities and tools emerging every month. This book is not a comprehensive guide, nor does it attempt to cover everything AI can do. It is a beginner's introduction, designed to help you take your first faithful steps into this new terrain.

Think of it as a starting point, not a final word. You will not find every answer here—but you will find direction, encouragement, and practical tools to begin using AI in ways that honor your calling and serve your community.

The journey ahead is full of discovery. Let this book help you get started.

An Invitation to Stewardship

I invite you to approach this journey with curiosity and caution. Come with open hands, ready to learn, and a heart anchored in Christ.

You do not have to become an expert. You do not need to master every prompt. But you do need to take the next faithful step. God is already at work in your community, and He may be calling you to steward new tools in new ways to serve His timeless mission.

As you read, pause often. Pray. Reflect. Ask the Spirit:

- *How can this serve the people You have called me to shepherd or lead?*

- *How can this multiply, rather than distract from, the ministry You have entrusted to me?*

At the end of the day, this book is just as much about faithful stewardship than it is about AI. It is about using what God has placed in your hands wisely, creatively, and in alignment with His Word.

So, let us begin: *together*.

How to Use This Book

This book is written especially for those who hesitate at the thought of blending faith and technology. If you've ever wondered, *"Can AI really fit with the things of God?"* — this book is for you. It is not a replacement for prayer or the Spirit, but a companion for the journey, a manual for ministry, and, if you allow it, a mirror reflecting the work God has entrusted to you.

Read With Open Hands and an Open Spirit

Do not come simply looking for tech tips or quick fixes. Come expecting revelation. Every chapter offers practical tools, but behind the tools lies a greater invitation: to see how God can use this moment in history to multiply the work of His church.

Approach these pages prayerfully. As you read, ask: *"Lord, what are You showing me for my house, my ministry, and my community?"*

Rooted in Scripture

Scripture is woven throughout, because God's Word *still* speaks. It anchors every innovation in eternal truth, reminding us that while tools and technologies may shift with the times, the voice of God remains our steady guide.

I use the terms AI and ChatGPT interchangeably. However, there is a difference.

AI vs. ChatGPT

- **Artificial Intelligence (AI)** is the broader field. It describes computer systems that can imitate aspects of human learning, reasoning, and communication. Think of it as the "big picture" of technology.

- **ChatGPT** is just one expression of AI. Created by OpenAI, it is a tool that uses natural language to interact with you — answering questions, generating ideas, and assisting with tasks.

Think of it this way: AI is like *transportation*, and ChatGPT is like *a car*. Transportation can take many forms — planes, trains, bicycles — but a car is one specific way to get around. In the same way, AI covers many possibilities, while ChatGPT is one particular way of experiencing it.

Why ChatGPT in This Book?

There are many AI tools available today - including Claude, Gemini, and Perplexity - and new ones seem to appear almost every month. For the sake of clarity, in this book I refer mainly to ChatGPT. Using one consistent platform gives you a clear starting point to practice with.

The prompts and principles I share, however, are not limited to ChatGPT. Most AI platforms work in a similar way: you give them a prompt, and they respond. The point is not *which* tool you use, but *how* you use it - with wisdom, discernment, and an ear tuned to God's voice.

Get Started Practically

If you want to follow along, you can create a free account at chatgpt.com. Try the examples, test out the prompts, and notice how the tools respond. The best way to learn is by doing - prayerfully, thoughtfully, and with a heart open to God's leading.

Move Around the Book as Needed

This is not a linear textbook. If you are focused with administration, read that chapter. If you are seeking new ways to serve the streets, go to T*he Streets We Serve*. If you are praying through how to fund vision and grant writing, turn to the chapters on *Stewardship*. Each section stands alone but also connects back to the larger vision of building a future ready church.

Gather a Team and Read Together

Do not keep this to yourself. Pull in your staff, your deacons, your youth leaders, and your lay volunteers. Imagine the power of a Wednesday night workshop where you read a chapter together and then brainstorm how it applies to your ministry. The Kingdom was never meant to be carried by one set of shoulders.

Look for the Marthas

As you read, ask: *"Who in my church is carrying a weight that technology could help lift?"* It may be the church secretary juggling announcements, the outreach leader tracking events by hand, or the media volunteer editing late into the night. This book will give you ways to release them. That release opens space for more prayer, more family time, and more rest.

Always Test by the Spirit

Technology can strengthen ministry, but it must never replace prayer, discernment, or wisdom. As you try new things, keep your ear to the ground and your heart tuned to heaven. The Spirit is still the best guide we have.

Dream Beyond the Walls

Some chapters will call you to see ministry outside the sanctuary. That includes prisons, reentry programs, digital neighborhoods, and the streets our ancestors once walked. Let those chapters stir holy imagination about what is next for your church in a world that is rapidly shifting.

A Final Word of Encouragement

This book is *not* about replacing people or spiritual authority. It is about equipping you with wisdom and tools that can lighten your load and expand your ministry impact. Take what is useful. Adapt it to your context. Most importantly, keep Christ at the center of all you do.

In short, use this book the way you use a church mother's wisdom. Return to it often. Pull from it as needed. Allow it to both comfort and challenge you. And as you move through these pages, remember: *worship is greater than work.*

PART I - FOUNDATIONS & VISION

Chapter 1 – Why This? Why Now?

STANDING AT THE CROSSROADS OF TOMORROW

> *"From the tribe of Issachar, there were 200 leaders... All these men understood the signs of the times and knew the best course for Israel to take."*
> — 1 Chronicles 12:32 (NLT)

This Is a Defining Moment

We are living in a moment of rapid change. Technology is evolving faster than most of us can process. Artificial Intelligence is no longer something from the future. It is something that is already here, woven into daily life, conversation, and culture.

Pastors and ministry leaders stand at a critical crossroads. The question is not whether AI will impact the world. It already has. The real question is whether we will approach this new reality with wisdom, courage, and vision, or shrink back in confusion and fear.

History shows us that the Church has always had a choice during moments of innovation. We can retreat and resist, or we can listen to the Holy Spirit and engage with clarity and purpose. The leaders from the tribe of Issachar were praised because they understood the times and knew what to do. That is what we are being invited into now.

Why the Church Cannot Stay Silent

Artificial Intelligence is shaping how people work, think, learn, and connect. It is showing up in schools, workplaces, homes, and even in conversations about faith. Young people are already using it. Church members are asking about it. Ministry teams are experimenting with it. And the world is watching how the Church will respond.

This is not just a technical conversation. It is a discipleship issue. It is a leadership issue. It is a stewardship issue.

To avoid the conversation is to leave people without guidance. To dismiss it as a trend is to miss an opportunity. To engage it prayerfully and wisely is to lead with both conviction and compassion.

What Is at Stake

If we do not understand the tools of the age, we risk becoming disconnected from the people we are called to reach and serve. If we do not speak into these cultural shifts, someone else will.

But if we take the time to understand, reflect, and lead through it, we position the Church to thrive, not just survive.

This is not about chasing relevance. It is about remaining faithful in a changing world.

Faithfulness in a New Landscape

Faithfulness does not mean keeping things exactly as they were. Faithfulness means walking closely with God as things change around us. It means listening for His direction in unfamiliar terrain. The early church adapted under the guidance of the Spirit. So can we.

God has not changed. The gospel has not changed. But the tools in our hands have.

If we learn how to use these tools with discernment and integrity, we can lighten administrative burdens, extend our reach, and support the sacred work of forming people in Christ.

A Word to the Hesitant Leader

You do not need to know all the jargon. You simply need to be open to learning and willing to lead from a place of wisdom.

Your voice matters. Your discernment matters. Your presence matters more than any tool ever will.

This journey is about strengthening what matters most. It is about helping you build capacity so you can stay present with God and with people.

Looking Ahead Together

Throughout this guide, you will learn how to think about AI biblically, how to use it practically, and how to lead others through it pastorally. You will be equipped not just to adapt but to disciple in the age of AI.

We are standing at the crossroads. This is not a moment for fear. This is a moment for faith.

Let us move forward with open hearts, clear minds, and deep trust in the One who knows the end from the beginning.

Chapter 2 – Seeing Clearly: What AI Is and What It Is Not

UNDERSTANDING CHATGPT WITH WISDOM AND CAUTION

> *"The prudent see danger and take refuge, but the simple keep going and pay the penalty."*
> — Proverbs 22:3 (NIV)

Clarity Before Use

When it comes to Artificial Intelligence (AI), especially tools like ChatGPT, the greatest danger is not in using it. The greater danger is using it without clarity. As leaders, we must not be driven by fear or hype. We must be guided by wisdom.

AI is a powerful tool, but it is not a person. It can assist in ministry, but it should never define it. In the same way that a hammer can build a house or do damage depending on the hand that holds it, AI must be used with discernment.

Before we talk about how to use AI in ministry, we must first understand what it is, what it is not, and what it can never be.

What AI Is

AI is trained on vast amounts of text and data to generate human-like responses. It can help with tasks like writing, summarizing, organizing, translating, and brainstorming. It can also help us save time, simplify processes, and support communication across teams.

Think of it like a virtual assistant. It is good at:

- Drafting written content
- Offering suggestions and templates
- Organizing to-do lists or project steps
- Generating ideas for planning or problem-solving
- Assisting with repetitive tasks like scheduling or formatting

Used well, AI can free up mental and emotional space so leaders can focus more on people, presence, and purpose.

What AI Is Not

AI is not a mind, not a soul, and not a substitute for spiritual leadership. It does not pray. It does not discern. It does not carry conviction or revelation. And it certainly does not carry the Holy Spirit.

AI is not wise. It is not relational. It is not safe to trust with confidential information, sensitive pastoral matters, or your voice of spiritual authority.

AI is also not perfect. It makes mistakes. Sometimes it gives wrong information or misuses Scripture. It has no lived experience and no moral compass unless one is given to it by the user.

Treat it like a helpful assistant, not like a mentor. Use it to serve your calling, not to shape it.

What AI Will Never Be

AI will never be a pastor. It will never carry the burdens of your people. It will never pray with the grieving, rejoice with the hopeful, or weep with those who weep. It can write a sympathy card, but it cannot offer your presence. It can organize a discipleship plan, but it cannot walk alongside someone's process.

However, you, led by the Spirit, can do that.

Discernment Over Fear

The goal of this conversation is not to make AI the center of your ministry. The goal is to help you approach it with confidence, so that when you use it, you are still being led by the Spirit and rooted in truth.

Some people fear AI and avoid it altogether. Others rush into it and overuse it, letting it shape their ministry too much. Both responses are short-sighted. What God wants is discernment.

Discernment asks:

- Does this tool serve my assignment?
- Am I staying faithful to my message and calling?
- Is this helping me love people better, **not just do more?**
- Am I giving the tool too much power or voice?

Questions to Reflect On

- Do I truly understand what AI is and how it works?
- What fears or assumptions am I carrying about it?
- Have I asked the Lord how I should engage or not engage with this tool?
- Where could it save time or reduce stress without compromising care?

A Final Reminder

You are called to lead, not just manage. You are called to shepherd, not to automate. Technology can help you move faster, but only God can show you the direction. As you explore AI, keep your heart rooted in the Word, your ear tuned to the Spirit, and your eyes fixed on the One who called you.

Clarity leads to confidence. Use the tool, but stay in step with the Spirit.

PART II - LEADING THE CHURCH WELL

Chapter 3 – Putting the House in Order

ALIGNING ADMINISTRATION WITH THE HEART OF MINISTRY

> *"But everything should be done in a fitting and orderly way."*
> — 1 Corinthians 14:40 (NIV)

Why Structure Matters in Ministry

God is a God of order. From creation to the early church, Scripture shows us that when the Spirit moves, there is form and structure. It is not unspiritual to have a plan. It is unwise not to.

Administration is not the opposite of anointing. In fact, it is what helps sustain the move of God over time. Without clarity and order, people get confused, things fall apart, and leadership becomes reactive instead of purposeful.

Many pastors and leaders are faithful in vision but struggling in execution. The reason is often simple. There are too few systems, or the ones that exist are outdated, unclear, or only live in someone's head.

The Wisdom of Jethro: Shared Leadership Requires Systems

In Exodus 18, Moses was leading with full passion but poor structure. He was handling everything himself until his father-in-law, Jethro, said something honest and loving:

> *"What you are doing is not good. You and these people who come to you will only wear yourselves out."*
> — Exodus 18:17-18 (NIV)

This is still true. Ministry without administrative systems will eventually wear out the leader and the people they are trying to serve.

The solution Jethro gave was simple: delegate responsibility, establish a structure, and create levels of support so Moses could focus on what only he could do.

Administration Is an Expression of Care

Order is not about control. It is about care. Clear systems help people feel safe, seen, and supported. Some examples include:

- A check-in system that helps parents trust the children's ministry
- Volunteer schedules that prevent burnout and confusion
- Budget tracking that ensures financial integrity
- A calendar that reflects both margin and mission

Paul told Titus,

> *"The reason I left you in Crete was that you might put in order what was left unfinished."* — Titus 1:5 (NIV)

The unfinished work of many ministries is not inspiration, but implementation. Order finishes what vision starts.

Using AI Tools to Strengthen Administration

Administrative work often involves repeatable, time-consuming tasks. While it is important to remain personally involved, you do not have to carry the load alone. Technology can help.

Tools like ChatGPT can assist with:

- Creating meeting agendas or summaries
- Developing monthly checklists
- Writing emails or announcements
- Designing onboarding guides
- Building annual calendars

These tasks do not require deep spiritual insight, but they do require time and consistency. When you use tools wisely, you free up your energy for prayer, relationship, and strategic leadership.

Practical ChatGPT Prompt Examples

Here are some specific ways you can use AI tools like ChatGPT to support your systems:

Meeting Notes to Action Plan

> **Prompt Example:**
> Turn these staff meeting notes into a to-do list with names, deadlines, and categories. *(Paste notes here)*

Monthly Operations Checklist

> **Prompt Example:**
> Create a monthly task list for church operations. Include facility checks, supply inventory, event planning, and follow-ups.

Volunteer Rotation Schedule

> **Prompt Example:**
> Create a volunteer schedule for Sunday teams with 10 people rotating across 4 roles. Ensure no one serves more than 2 times per month.

Administrative Calendar

> **Prompt Example:**
> Build a 12-month planning calendar for a church with key administrative tasks, seasonal outreach, and team planning checkpoints.

Follow-Up Templates

> **Prompt Example:**
> `Write a friendly email reminder asking team leads to submit their quarterly goals by Friday.`

Common Organizational Gaps and How to Fix Them

Problem	System Solution	Prompt Example
Missed tasks or unclear follow-up	Shared task tracker	Create a template for tracking team tasks and deadlines.
Volunteers showing up late or unprepared	Serve team schedule with reminders	Build a Sunday morning volunteer call time and responsibilities chart.
Overwhelm with recurring tasks	Checklist with weekly and monthly tasks	Make a repeating checklist for weekly and monthly admin responsibilities.

Even small improvements bring peace and predictability to your team.

A Word of Warning: Movement Without Direction

It is possible to be busy and still be disorganized. A packed calendar does not mean you are bearing fruit. More meetings do not equal more mission.

Ask yourself regularly:

- Are our systems helping us serve people better?
- Are we just staying active, or are we moving with purpose?

Avoid letting tools become your master. They are there to serve the mission, not replace your discernment.

> *"Commit to the Lord whatever you do, and He will establish your plans."*
> — Proverbs 16:3 (NIV)

Reflection Questions: Is Your House in Order?

- What part of your ministry feels most disorganized or reactive?
- Which repeated task could be made easier with a better system or template?
- Where are you still doing things manually that could be automated or delegated?
- What would change if your admin systems supported peace instead of pressure?
- Who else can help you build and maintain systems that reflect your values?

Order Makes Room for Presence

The early church grew not only because of prayer and power but because of shared responsibilities and clear roles. When the apostles realized they were neglecting some areas of care, they responded by appointing others and organizing for growth.

Order does not quench the Spirit. It welcomes Him.

Bringing your administrative house into order is not a distraction from ministry. It is part of it. Systems should serve people, not the other way around. When done well, they create space for rest, renewal, and the move of God.

Put the house in order, not just for efficiency, but so that the people of God can be empowered as they serve.

Chapter 4 – Caring for the Flock

MEMBER CARE AND COMMUNICATION IN A DIGITAL WORLD

> *"Be shepherds of God's flock that is under your care, watching over them... not lording it over those entrusted to you, but being examples to the flock."*
> — 1 Peter 5:2-3 (NIV)

The Heart of Shepherding

Caring for people is one of the most sacred responsibilities in ministry. At the center of pastoral leadership is the call to shepherd. This means walking with people, praying for them, being present in their struggles, celebrating their joys, and guiding them in their journey with Christ.

It is not about perfect programs or polished performances. It is about presence, love, and trust. Real care cannot be automated. But it can be supported by thoughtful systems and timely communication.

Why Member Care Often Gets Overlooked

Most pastors and leaders care deeply. The challenge is not desire. The challenge is capacity. Between weekly services, team meetings, crises, and administration, it is easy to lose track of who needs a phone call, who is grieving, or who has not been seen in a while.

Without intentional structure, people can slip through the cracks. Not because they are unloved, but because the load is too heavy and the system is too loose.

That is where tools, technology, and shared care can make a difference. They should not replace the shepherd's heart, but to help protect it.

Using AI to Strengthen (Not Replace) Care

Artificial Intelligence, when used with wisdom and boundaries, can serve your care ministry by helping you stay organized, consistent, and proactive. It cannot comfort the brokenhearted, but it can help you remember to call them. It cannot replace prayer, but it can draft the message that opens the door to it.

Here are a few ways AI can assist your pastoral care efforts:

- Create follow-up messages for hospital visits or prayer requests
- Draft reminders to check in after a loss or surgery
- Build care lists and task trackers
- Generate birthday or anniversary messages
- Organize volunteer care team schedules
- Help write devotionals or encouragement notes for members

These small touches, when offered with love, create a culture where people feel seen and valued.

Sample ChatGPT Prompts for Member Care

Follow-Up Message After a Crisis

> **Prompt Example:**
> ```
> Write a kind message to follow up with a
> church member who recently lost a loved one.
> Keep it personal and sensitive.
> ```

Check-In for Absent Members

> **Prompt Example:**
> ```
> Create a gentle message to check in with
> someone who has not attended church in the
> past month.
> ```

Monthly Member Care Calendar

Prompt Example:
```
Make a monthly schedule for checking in on
members based on birthdays, prayer needs, and
pastoral care lists.
```

Encouragement Notes

Prompt Example:
```
Write 3 short encouragement messages based on
Psalm 23 for people going through hard times.
```

Communication That Builds Trust

In a digital world, communication is care. When people feel informed and included, they are more likely to stay connected and engaged. When people feel forgotten or confused, they tend to drift away quietly.

Other ways that AI can help streamline communication include:

- Writing ministry announcements
- Drafting reminder texts or emails
- Translating messages for multi-language communities
- Formatting updates for newsletters or bulletins
- Creating welcome messages for new guests

These are not flashy tasks, but they are essential. Good communication honors people's time and builds trust across the church.

A Word of Caution

Do not over-rely on AI to do relational work. Use wisdom. Never input private prayer requests, counseling notes, or personal stories into AI tools. Protect the privacy and dignity of your members.

Always review and personalize what the tool generates. Let your own heart, theology and instinct lead the final message.

Technology is here to support ministry, not to replace it.

Questions to Reflect On

- Are people in our church being followed up with in meaningful ways?
- Who is responsible for care, and are they supported?
- What systems could make sure no one is forgotten?
- How can we communicate more clearly and consistently?

Love in Every Detail

Caring for the flock is not always dramatic. Often, it is the quiet phone call, the remembered birthday, or the timely text that makes someone feel truly loved. These small actions reflect the Shepherd's heart.

When you pair spiritual attentiveness with thoughtful tools, you multiply care without losing compassion. You stay present without becoming overwhelmed. You create a culture where people feel valued, not just managed.

Let love guide your communication. Let wisdom guide your tools. Let the Spirit guide your heart.

Chapter 5 – Teaching That Transforms
DISCIPLESHIP & EDUCATION

> *"Then Jesus came to them and said, 'All authority in heaven and on earth has been given to me. Therefore go and make disciples of all nations... teaching them to obey everything I have commanded you.'"*
> — Matthew 28:18-20 (NIV)

Teaching Is More Than Content

Teaching has always been at the center of the Church's mission. Jesus did not just call us to inspire crowds or fill rooms. He called us to make disciples. That means forming hearts, shaping minds, and guiding lives toward obedience and transformation.

Real teaching is not about transferring information. It is about forming character and cultivating Christlikeness. The classroom of discipleship is not limited to Sunday services or midweek Bible study. It includes conversations, mentorship, daily habits, and the flow of spiritual life together.

In a time where content is everywhere, the role of the teacher must become more intentional, more personal, and more focused on transformation over performance.

The Opportunity of the AI Age

With AI tools and platforms, the process of creating, organizing, and delivering educational resources is becoming easier. Leaders can now access templates, outlines, and ideas in moments that used to take hours.

But speed is not the goal. Depth is.

AI can help us create more consistent and well-structured materials. It can offer support for planning and teaching. What it cannot do is lead someone into obedience or spiritual maturity. That still requires time, relationship, and the work of the Holy Spirit.

When used wisely, AI can support discipleship by:

- Drafting devotional content or study guides
- Offering reflection questions based on Scripture
- Suggesting learning pathways for new believers
- Helping leaders plan teaching calendars
- Creating group discussion starters

These tools should not be the teacher. They should serve the teacher.

Practical Ways to Use AI in Discipleship

Here are a few examples of how a pastor or leader might use ChatGPT to support educational goals:

Discipleship Journey for New Believers

> **Prompt Example:**
> ```
> Create a 6-week discipleship path for new believers that includes weekly Scripture, one spiritual habit, and a reflection question.
> ```

Small Group Curriculum Outline

> **Prompt Example:**
> ```
> Draft a 4-week Bible study on the fruit of the Spirit for a small group. Include main passages, key insights, and a practical challenge.
> ```

Youth or Children's Ministry Devotional

Prompt Example:
```
Write a one-page devotional on kindness for
kids aged 9-12. Use simple language and
include a memory verse.
```

Teaching Calendar

Prompt Example:
```
Create a 12-month teaching plan for adult
Bible classes based on themes like prayer,
identity in Christ, and spiritual maturity.
```

These examples show how AI can provide structure and creative input, while leaving space for your voice, your theology, and your insight to remain central.

The Heart of the Teacher Must Lead

Teaching that transforms does not come from well-polished outlines. It comes from a life surrendered to Christ, a love for people, and a clear sense of calling.

Do not let technology do what only the Spirit can do. Let AI help you prepare. Let God shape the outcome.

It is still your voice, your heart, your example, and your relationship with the learner that produces fruit over time.

A Word of Caution

AI cannot replace mentorship, community, or correction. It cannot be trusted to rightly divide the Word of truth without your careful review.

Do not copy and paste AI-generated teaching without spiritual discernment. Let it be a tool, not a shortcut. Let it enhance what you already know God is asking you to teach.

Reflection Questions for Teachers and Disciple-Makers

- What is the goal of my teaching?
- Am I creating room for transformation, not just transfer of information?
- Where could structure or creativity help me disciple more effectively?
- How can I remain faithful to the Word while exploring new tools?

Teaching in the Spirit, With Help from the Tools

Jesus taught in fields, homes, synagogues, and by the sea. He met people where they were and invited them into a new way of living.

Today, the classroom has expanded. People learn through podcasts, apps, texts, and online courses. But the mission has not changed.

AI can help us plan and deliver teaching more efficiently. But only the Spirit can transform hearts.

Let your teaching be Spirit-filled, biblically sound, and made stronger by the support of new tools. Let your words lead people not just to learn, but to live the truth.

Chapter 6 – Shepherding the Next Generation
AI IN YOUTH AND CHILDREN'S MINISTRY

> *"Let the little children come to Me, and do not hinder them, for the kingdom of heaven belongs to such as these."*
> — *Matthew 19:14 (NIV)*

Why Youth and Children's Ministry Needs Fresh Tools

Children and teenagers today are growing up in a world where technology is second nature. They learn, play, and connect through screens and digital platforms. As the Church, we have the opportunity to meet them where they are - not to entertain, but to engage, disciple, and equip.

Artificial intelligence is not a replacement for relationships, mentorship, or human connection. But it can become a powerful tool in helping churches create experiences that are relevant, meaningful, and spiritually formative. It can help you connect more deeply, plan more effectively, and lead more intentionally.

Youth leaders and children's ministry teams are often stretched thin. AI can lighten the load and multiply impact when used with wisdom and care.

Practical Ways AI Can Support Youth and Children's Ministry

1. Lesson Planning Made Simple

AI tools like ChatGPT can help create Bible study outlines, youth group devotionals, and children's lessons tailored to specific age groups or themes.

> **Prompt Example:**
> Create a 30-minute Bible study for middle school students on the topic of courage using the story of David and Goliath.

2. Interactive Learning Activities

From games and icebreakers to role-playing scenarios and creative storytelling, AI can generate fresh ideas that make learning fun and interactive.

> **Prompt Example:**
> Suggest 5 Bible-based team games for a youth retreat focused on friendship and unity.

3. Content for Parents and Families

AI can assist in writing weekly summaries, parent updates, or discipleship tips for families to continue the conversation at home.

> **Prompt Example:**
> Write a short newsletter blurb for parents recapping this week's lesson on forgiveness for 3rd-5th graders.

4. Social Media and Engagement

Keep youth connected with encouraging posts, devotional texts, or announcements formatted for Instagram, TikTok, or email.

> **Prompt Example:**
> Draft an encouraging 1-paragraph devotional for teens about trusting God during exam season.

5. Volunteer Support and Training

Create onboarding guides, training outlines, and team meeting agendas to support youth volunteers and ministry leaders.

> **Prompt Example:**
> Write a volunteer orientation guide for new high school small group leaders, including expectations and best practices.

6. Event Planning Help

AI can generate checklists, timelines, or creative themes for youth camps, VBS, or back-to-school nights.

> **Prompt Example:**
> List 10 creative and gospel-centered themes for a one-day youth conference.

7. Student Discipleship Pathways

Support individualized growth by creating age-appropriate reading plans, reflection questions, or goal-setting guides for spiritual formation.

> **Prompt Example:**
> Create a 4-week devotional guide for high schoolers about discovering their spiritual gifts.

Guarding What Matters Most

The heart of youth and children's ministry is presence, trust, and mentorship. We should not rely on AI to pray with a teen, sit with a child who is grieving, or guide a young person through personal struggles. Those moments must always remain human and Spirit-led.

With Hope for the Future

The next generation is not waiting for the Church to catch up. They are forming ideas about faith, identity, and truth right now. AI can help us speak their language, reach their hearts, and design spaces where they feel seen, known, and loved.

You do not have to be a tech expert to begin. You simply have to be willing to explore. With a teachable spirit and a pastor's heart, you can shepherd this generation into deeper faith using every tool available—including AI.

Let the children come. And let the Church be ready.

Chapter 7 – The Streets We Serve

COMMUNITY ENGAGEMENT

> *"And the Lord said to him, 'Go, for he is a chosen instrument of mine to carry my name before the Gentiles and kings and the children of Israel.'"*
> — Acts 9:15 (NIV)

Ministry Goes Beyond the Sanctuary

The mission field begins at your doorstep. Community engagement is not an optional ministry. It is the heart of the gospel. Jesus walked through cities, villages, and open fields meeting people in their places of need. Today, we are called to do the same. Whether through reentry programs, feeding ministries, advocacy, or local partnerships, the Church must be visible and active in the lives of real people.

Digital tools, including Artificial Intelligence, are no substitute for presence and compassion. But when used well, they can multiply your capacity, improve your communication, and help you lead with strategy and clarity.

The Needs Are Great. So Is the Opportunity.

Across cities and towns, people are facing challenges like homelessness, food insecurity, reentry after incarceration, addiction recovery, unemployment, and lack of affordable housing. The Church has always played a role in responding to these needs. But modern problems often require modern tools.

This is where ChatGPT and other AI systems can offer real support. They can help churches prepare, plan, communicate, and collaborate more effectively with their communities.

How AI Can Support Community Outreach

Here are ways AI can help you serve more effectively in specific outreach areas:

1. Reentry Programs (Returning Citizens)

Churches that support individuals returning from incarceration play a powerful role in reducing recidivism and restoring dignity. These ministries often need structure, training, and encouragement.

You can use AI to help you:

- Draft reentry program outlines or curriculum for life skills and discipleship
- Write letters of encouragement or welcome packets for returning citizens
- Create job readiness workshop materials
- Suggest devotional topics focused on restoration and identity
- Write grant proposal drafts for funding reentry support

Prompt Examples:

```
Create a six-week life skills course outline
for returning citizens.
```

```
Draft a welcome letter to someone being
released from prison and entering our church's
reentry program.
```

```
Help write job interview tips for people with
a criminal record.
```

Prompt Examples:

```
Write an invitation to our church's free
school supply giveaway.
```

```
Translate a volunteer flyer into Spanish and
Haitian Creole.
```

```
Help design a neighborhood survey asking what
outreach efforts are most needed.
```

```
Create a list of questions for a prayer walk
through our community.
```

A Ministry That Knows Its Streets

Jesus walked the dusty roads of towns and villages, healing, feeding, and teaching. He knew the names of people and the pain of places. Today, churches must continue the outreach. AI can help you know your streets better, organize your work more effectively, and respond to needs with greater precision.

Technology should never take the place of love. But it can help your love go further. It can help you see patterns, organize volunteers, and communicate with clarity. When used with care, AI becomes a tool for outreach, not a replacement for relationship.

Presence Still Matters Most

The Church must remain rooted in the real world, even as it explores digital tools. The greatest gift you give is not a form, a flyer, or a program. It is your presence. It is the consistency of showing up, knowing names, and offering grace without condition.

> List Bible verses and reflections for a
> reentry small group study.

2. Soup Kitchens and Meal Distribution

Feeding ministries require logistics, volunteer coordination, and ongoing communication.

You can use AI to help you:

- Write volunteer instructions and shift schedules
- Create flyers for food distribution events
- Draft emails or texts to update teams and guests
- Plan monthly menus with nutritional balance
- Draft donor letters or thank-you notes

Prompt Examples:

> Write a flyer inviting the community to our free Friday soup kitchen.

> Create a checklist for volunteers serving meals to homeless guests.

> Draft a thank-you letter for a local business that donated food supplies.

> Help plan a rotating four-week meal schedule for 75 people.

3. Homeless Outreach

Serving people without housing requires sensitivity, coordination, and safety planning.

You can use AI to help you:

- Create emergency response plans for cold weather or heatwaves
- Draft outreach team training materials
- Write devotionals or encouragement notes to hand out
- Design resource guides with shelter, clinic, and food locations
- Summarize best practices for trauma-informed care

Prompt Examples:

```
Write a devotional card of encouragement for
someone experiencing homelessness.
```

```
List basic supplies we should include in
winter care packages.
```

```
Help draft a guide for outreach volunteers
working in homeless shelters.
```

```
Create a printable sheet listing local shelter
and clinic contacts.
```

4. Engaging with City Council and Local Leaders

Effective outreach often requires public partnerships and advocacy. Churches that build bridges with local government can help influence policy, secure resources, and create long-term change.

You can use AI to help you:

- Draft letters to city officials or council members
- Prepare talking points for public meetings
- Summarize local policy proposals or community needs
- Write press releases or public statements
- Help structure community presentations or forums

Prompt Examples:

```
Draft a letter to the city council requesting
support for a church-led reentry program.
```

```
Summarize the key points of a housing proposal
into plain language for our congregation.
```

```
Create an outline for a church-hosted town
hall on youth violence prevention.
```

```
Write a statement from a pastor supporting
local efforts to increase shelter funding.
```

5. Partnership with Local Schools

Churches and schools can form meaningful partnerships that support students, families, and educators. Whether through mentorship, tutoring, school supplies, or special events, the Church can be a steady and caring presence.

You can use AI to help you:

- Draft proposals for collaboration with school leaders
- Create parent communication materials

- Plan school supply drives or volunteer reading days
- Develop mentoring and afterschool program outlines
- Write devotionals or encouragement notes for teachers

Prompt Examples:

```
Write a proposal for a church-school
partnership focused on mentoring.
```

```
Create a flyer for our back-to-school supply
giveaway.
```

```
Draft a letter to encourage public school
teachers in our neighborhood.
```

```
List ideas for volunteer roles to support
students and families.
```

6. General Community Engagement

Beyond specific ministries, churches are called to know and love their neighbors. AI can help churches communicate clearly, respond quickly, and stay organized.

You can use AI to help you:

- Write outreach emails, announcements, or text campaigns
- Translate content into different languages for diverse communities
- Brainstorm ideas for block parties, health fairs, or back-to-school drives
- Draft prayer guides for city transformation
- Organize community needs surveys or feedback forms

Let AI serve your mission, not shape it. Let it help you organize, plan, write, and extend—but let it be the hands and feet of Christ that truly touch the streets you serve.

Chapter 8 – A Church for All: Accessibility, Inclusion, and AI

LANGUAGE, DISABILITY, AND THE GOSPEL WITHOUT BARRIERS

> *"Then how is it that each of us hears them in our native language?"*
> — Acts 2:8 (NIV)

The Gospel Is for Everyone

From the day of Pentecost, when people from many nations heard the gospel in their own languages, the message of Jesus has always been one of accessibility. The Church is called to be a place where everyone—regardless of language, ability, or background—can encounter the presence of God and find belonging.

But in practice, accessibility can be challenging. Many churches struggle to serve people with hearing impairments, language barriers, or learning differences. AI offers new ways to bridge these gaps with compassion, innovation, and care.

Why Accessibility Matters in the Church

Accessibility is not simply a matter of technology. It is a matter of theology. When we remove barriers to communication and participation, we reflect the heart of God, who welcomes all to His table. Making our ministries more inclusive honors the image of God in every person.

Today, churches are more diverse than ever. Immigrant families, second-language speakers, neurodiverse children, and aging members all bring unique needs. AI can help churches respond with grace and clarity.

How AI Supports Language Diversity

1. Real-Time Translation Tools

AI-powered translation is now faster and more accurate than ever. Tools like Google Translate, DeepL, and Microsoft Translator allow churches to:

- Translate sermons and notes into multiple languages.
- Offer real-time captions during livestreams or in-person services.
- Communicate clearly in multilingual congregations.

New Technology Highlight

AI Translation Headphones allow real-time multilingual communication.

- A pastor could preach in English while attendees hear it in Spanish or French through earbuds.
- Volunteers can welcome newcomers even without speaking their language.

> **Prompt Example:**
> ```
> Translate this welcome message into Spanish, French, and Swahili in a warm and respectful tone.
> ```

If you use AI to translate into another language, it is best to review the output. Whenever possible, confirm the accuracy with a human.

2. Multilingual Outreach Content

ChatGPT can help write outreach flyers, devotionals, or social posts in multiple languages with cultural sensitivity. This is especially useful for multicultural events or ministries that serve immigrant populations.

AI and Disability Inclusion

1. Audio and Visual Accessibility

AI tools can convert spoken words into text, and text into speech. This allows for:

- Closed captions for livestreams or sermon videos
- Read-aloud features for visually impaired members
- Voice-to-text sermon notes for those who struggle with manual writing

Tools That Help:

- Otter.ai: Transcribes sermons and classes in real time
- Microsoft Azure Cognitive Services: Offers speech-to-text, text-to-speech, and accessible navigation tools
- Be My Eyes (AI mode): Uses AI to describe images or surroundings to visually impaired users

Prompt Example:
```
Summarize this sermon transcript into simple
language for someone with a cognitive
disability.
```

2. Neurodiverse Support

Some members may have learning disabilities, ADHD, or autism. AI can help you:

- Break down complex teaching into simple steps
- Reformat written materials into more visual, readable content
- Personalize devotionals to suit different learning styles

3. Sign Language and Visual Aids

While AI cannot yet replace human interpreters, it can assist by:

- Suggesting visuals, symbols, or simplified scripts for children or the deaf community
- Providing summaries of worship content for pre-service access

Leading with Intentionality

Using AI for accessibility is not just about being high-tech. It is about being thoughtful. Here are a few practices to guide your efforts:

- Ask and listen: Talk to members with different needs. Learn what would help them feel included.
- Test and improve: Try new tools, get feedback, and adjust over time.
- Honor dignity: Always protect people's privacy and avoid using assistive tools in ways that feel intrusive or demeaning.

The Body Needs Every Part

Paul reminds us in 1 Corinthians 12 that the body of Christ needs every part to function well. Accessibility is not a favor we extend—it is a reflection of our belief that every person has something valuable to offer the Church.

AI cannot replace love, but it can help express it more clearly. By breaking down language barriers and meeting people in their needs, churches become places where no one is left out and all are welcomed in.

Let us build ministries where everyone can hear, see, understand, and belong—together.

PART III – STEWARDSHIP OF TIME, TALENT & TREASURE

PART III – STOCKPILING OF TIME: TALENT'S TREASURE

Chapter 9 – Redeeming the Time
ADMINISTRATION & EFFICIENCY

> *"Be very careful, then, how you live—not as unwise but as wise, making the most of every opportunity, because the days are evil."*
> — Ephesians 5:15-16 (NIV)

Time Is a Ministry Resource

In ministry, time is one of your most limited and valuable resources. You can always create more content, gather more volunteers, or plan more programs. But you cannot create more time.

That is why stewardship of time and tasks is not just a productivity issue. It is a spiritual discipline. It allows you to stay available to God and others, rather than constantly playing catch-up. It helps you keep your focus on people, not paperwork.

Administration is not always glamorous, but it is necessary.

Why Efficiency Matters in Ministry

Being efficient is not about speed. It is about purpose. It is about using energy and time on what truly matters. Many ministry leaders are overwhelmed, not because they are lazy, but because they are overloaded.

There are too many meetings, too many emails, too many small decisions that drain energy. Without systems, the urgent pushes out the important. You spend more time reacting than leading.

God calls us to walk wisely, not wastefully. That means learning to manage the work, not letting it manage us.

Where AI Can Help You Redeem the Time

AI is not just for content creation. It can help simplify and support administrative tasks that normally eat up time and energy. This does not mean handing everything over to a machine. It means letting a tool carry what it can so you are free to carry what matters most.

Here are areas where AI can help:

- Generating checklists and timelines for events
- Creating follow-up workflows for guests or team members
- Writing policies or volunteer guidelines
- Automating reminders or standard messages

These small gains add up. They help reduce mental clutter and decision fatigue.

Practical Prompts for Church Administration

Weekly Task List for a Pastor

> **Prompt Example:**
> Make a weekly task list for a lead pastor that includes prayer, planning, staff meetings, pastoral care, and family time.

Guest Follow-Up Email

> **Prompt Example:**
> Write a warm and welcoming follow-up email for a first-time church guest. Include a thank-you and an invitation to connect.

Event Planning Checklist

> **Prompt Example:**
> Create a checklist for planning a church outreach event, including promotion, volunteer roles, and follow-up steps.

Staff Meeting Agenda Template

> **Prompt Example:**
> ```
> Draft a simple and focused agenda for a 1-hour
> ministry team meeting. Include devotion,
> updates, and decision items.
> ```

You can edit and customize these as needed, saving time and maintaining quality communication.

The Goal Is Peace, Not Pressure

Efficiency in ministry is not about working faster just to do more. It is about creating breathing room for what matters. When the administrative side of ministry is healthy and organized, it becomes easier to lead from a place of peace instead of pressure.

AI can help you get ahead on your calendar, simplify your workflow, and prevent things from falling through the cracks. It can also help your team function better by giving clarity and structure to shared work.

Use Tools, But Keep Sabbath

In the pursuit of efficiency, do not lose sight of rest. You are not a machine. You may be a shepherd, a servant, and/or a child of God. Redeeming the time also means protecting space to be still, to listen, and to receive.

Technology is a gift, but Sabbath is a command. Let tools help you work wisely, but do not let them rob your soul of margin and quiet.

Reflection Questions

- Where is most of my time going each week?
- What tasks could be simplified, delegated, or automated?
- Am I consistently overworking, or am I building in rest?
- How can I help my team stay clear and focused without burnout?
- Am I using technology to serve ministry, or has it started to shape the way I measure success?

Final Encouragement

You are called to lead with wisdom, not just busyness. Redeeming the time is not about doing everything. It is about doing the right things with the right heart.

Let systems serve your calling. Let tools remove distractions. And let your time be a testimony that God is leading, not just your calendar.

INNOVATION, CREATIVITY & EMPOWERMENT

Chapter 10 – Singing the Lord's Song in a Digital Land

PHOTOS, VIDEOS, MUSIC AND MEDIA

> *"How shall we sing the Lord's song in a strange land?"*
> — Psalm 137:4 (KJV)

We Are Still Called to Sing

Every generation has found new ways to express the truth of God's Word and the beauty of His presence. Whether through handwritten scrolls or livestream worship, the calling to declare God's greatness remains the same. The setting may change, but the song continues.

Today, we serve in a world shaped by digital experiences. Photos, videos, music, and media are no longer extras. They are primary ways people absorb truth, feel connected, and stay engaged. Rather than resisting the shift, the Church is invited to step into it with wisdom and creativity.

We are not just called to keep up. We are called to redeem the space. We are still singing, just in a new land.

Media Ministry Is Message Ministry

Media is not a side project. It carries the message. A blurry photo, a poorly edited video, or a confusing graphic can block the message from landing well. On the other hand, a clear image or well-produced video can make a message more memorable and shareable.

Churches today are using media to:

- Celebrate testimonies
- Share Scripture visually
- Invite people to events
- Inspire worship beyond Sunday
- Connect with the next generation

You do not have to be a professional media producer. You simply need to see creativity as a form of communication and media as a tool for ministry.

Where AI Can Help with Media Ministry

AI tools like ChatGPT can support your creative work, even if they do not replace the human eye or the Spirit's leading. AI helps you plan, write, structure, and organize your media efforts, saving time and sparking ideas.

You can ask ChatGPT to:

- Write video scripts
- Draft social media captions
- Suggest worship themes or lyrics
- Create promotional copy
- Organize your media production calendar
- Brainstorm content for campaigns or holidays

The goal is not to sound robotic. The goal is to remove the mental clutter so you can create with focus and joy.

What ChatGPT Can and Cannot Do

At the time of this writing, ChatGPT can generate text and images (when the right version is enabled), but it does *not* create audio or video files directly. However, other AI tools do support music composition, voice generation, and video editing. These tools are emerging quickly, with new features appearing almost every month.

Even though ChatGPT does not create media files, it can help you work with those files in powerful ways:

For Images

- Generate ideas for sermon illustrations, social media graphics, or event flyers with short text prompts
- Brainstorm visual concepts and themes for art, banners, or church branding
- Collaborate with design tools like Canva or Adobe by writing polished descriptions for logos or slides

For Audio (Sound and Music)

- Transcribe uploaded sermons, podcasts, or voice memos into readable text
- Summarize long audio into highlights, sermon points, or quotes for newsletters
- Draft scripts for devotionals or podcasts that you can record later
- Suggest background music styles or write lyrics for worship songs

While ChatGPT cannot produce a sound file of a choir or band, it can help you write lyrics, outlines, and narration that you or your team will record.

For Video

- Transcribe sermon videos or testimonies so they can be repurposed into blog posts or devotionals
- Write video titles, captions, or YouTube descriptions to reach a wider audience
- Draft scripts for announcements, seasonal videos, or children's skits
- Provide storyboards or creative outlines for holiday plays, welcome videos, or outreach clips

For Document Files

- Analyze, summarize, rewrite, or extract key ideas from sermon notes, reports, or devotionals

- Provide feedback on clarity, tone, or organization for any ministry writing
- What ChatGPT cannot do: It cannot edit visual formatting or change file types (such as converting a PDF to a Word doc)

If you need to turn scanned pages or printed materials into editable documents, tools like Adobe Scan or OCR apps can convert them. Once in text form, ChatGPT can review or enhance them.

ChatGPT Is Your Creative Assistant

ChatGPT does not replace your creativity. It partners with it. Think of it as a creative assistant that helps shape the words, structure the outline, and sharpen the message before it is shared through video, music, or design.

Let it help you with the building blocks. Then you and your team bring it to life.

Prompt Examples:

```
Write a caption for a youth group photo that
emphasizes joy and spiritual growth.
```

```
Give me five title ideas for a video series on
biblical wisdom for modern life.
```

```
Create a short script for a 90-second
announcement video about our upcoming men's
retreat.
```

```
Suggest three lyrical lines for a worship
chorus on God's faithfulness.
```

Draft a visual theme description for a fall sermon series that feels warm and hopeful.

Final Encouragement

In this digital age, creativity is not optional. It is part of your witness. People may see your photo before they hear your sermon. They may watch your short video before they visit your church. They may read your lyrics before they open their Bible.

Use these tools with care, with prayer, and with purpose. Keep singing the Lord's song, even in unfamiliar territory. You are not performing. You are proclaiming.

And you are not alone. The Spirit leads you. The Church surrounds you. And now, even your tools can serve you more faithfully.

Chapter 11 – Stewardship, Grant Writing, and Economic Empowerment

FUNDING

> *"Moreover, it is required in stewards that one be found faithful."*
> — 1 Corinthians 4:2 (NKJV)

Resourcing the Vision

God often provides vision before provision. As spiritual leaders, we are entrusted not only with sharing the gospel but also with managing the resources that support the work of ministry. This includes funding, property, staff, partnerships, and new opportunities for economic empowerment.

In today's environment, churches have access to funding beyond tithes and offerings. Grants, philanthropic partnerships, and community support can fuel initiatives in outreach, education, youth development, mental health, and more. To access these resources, the Church must learn how to communicate its vision in language funders understand.

Why Funding Matters for Ministry

Money is not the mission, but it helps fuel the mission. Churches need sustainable funding for:

- Building maintenance or rental
- Outreach and benevolence programs
- Technology and media resources
- Staffing and volunteer support

- Curriculum and discipleship materials
- Community development or empowerment initiatives

Without funding, vision often remains unrealized. Good stewardship brings it to life and multiplies its reach.

Speaking Two Languages: Church and Community

Inside the Church, we speak with passion and faith. Words like "revival," "deliverance," or "watch night" hold deep spiritual meaning. But those same words may confuse someone reviewing a grant proposal.

This does not mean changing your beliefs. It means adapting your language for clarity. The Apostle Paul said he became all things to all people. Likewise, our ministry language must be flexible enough to reach new ears without compromising truth.

Why Language Matters in Funding

Faith-based programs often miss funding opportunities not because they lack value, but because they are not framed in the language of impact. "Spiritual growth" might be called "emotional wellness." "Discipleship" might align with "youth mentorship." Proper translation is not about hiding faith but building bridges.

Translating Common Church Programs

Many church programs already align with community goals and public funding priorities. Below are examples of how to describe them in broader, more fundable terms:

Vacation Bible School (VBS)

- Church Language: A summer program for children to learn Bible stories through music, crafts, and play.
- Fundable Framing: A youth enrichment camp providing educational programming, literacy support, and social-emotional development.
- Related Areas: Summer learning, youth development, food access, arts and culture.

Prayer Breakfast

- Church Language: A gathering for prayer, worship, and fellowship over breakfast.
- Fundable Framing: A wellness-focused community gathering promoting mental resilience and social connection.
- Related Areas: Mental health awareness, civic engagement, public health.

Watch Night Service

- Church Language: A New Year's Eve worship service focused on reflection and prayer.
- Fundable Framing: A year-end event for grief recovery, reflection, and trauma-informed goal setting.
- Related Areas: Mental health recovery, intergenerational wellness, cultural storytelling.

Revival Services

- Church Language: Multi-day services emphasizing repentance and renewal.
- Fundable Framing: Community gatherings using the arts to promote emotional healing and connection.
- Related Areas: Performing arts, mental health, substance recovery, violence prevention.

Youth Explosion

- Church Language: A high-energy event for youth engagement in worship.
- Fundable Framing: A youth leadership summit focused on creative expression and identity formation.
- Related Areas: Youth development, leadership, digital media, mental health.

Men/Women's Day

- Church Language: A celebration of men/women's contributions to the church.
- Fundable Framing: A men/women's empowerment event with a focus on wellness, leadership, and support services.
- Related Areas: Men/women's health, economic opportunity, trauma recovery.

Mental Health and Community Wellness: Hidden Connections

Many churches already provide mental and emotional support without labeling it that way. Programs like grief counseling, altar prayer, support groups, and wellness check-ins meet real mental health needs.

If your church offers:

- Pastoral counseling
- Prayer teams that follow up with the bereaved
- Monthly wellness or caregiver support gatherings
- Youth spaces for emotional expression
- Bible studies on resilience and identity

Then you already have the foundation for a mental health initiative. Translating these efforts into recognized terms opens doors to new partnerships and funding.

The Opportunity of Grant Writing

Grants are not just for large organizations. Thousands are available to support faith-based education, social services, arts, and community healing. Grant writing simply means presenting a real need, a thoughtful plan, and a clear budget.

Even without a background in professional writing, you can start with what you know. Combine that with support tools, and your ministry's vision can be shared with clarity and strength.

How ChatGPT Can Help with Grant Writing

ChatGPT is not a replacement for discernment or financial management. But it can help lighten the load when preparing proposals, reports, and presentations.

Ways ChatGPT Can Help:

- Draft sample narratives describing your church's mission
- Create proposal templates with goals and outcomes
- Suggest data or storytelling strategies to demonstrate impact
- Generate outlines for budgets or expense categories
- Edit and clarify writing for tone and professionalism

Sample Prompts for Grant Writing Support

```
Rewrite the following ministry description in
professional grant-application language:
[insert your description].
```

```
Create a mission statement for our church's
community kitchen.
```

```
List five measurable outcomes for a church-led
literacy project.
```

```
Draft a thank-you letter to a foundation for a
grant award.
```

```
Create an outline for a grant proposal for a
church-based mentoring program. Include
problem statement, solution, outcomes, and
evaluation methods.
```

These prompts can help you go from a blank page to a polished draft.

How ChatGPT Can Help Beyond Writing

Beyond grant writing, ChatGPT can also assist with communications and translation:

It Can:

- Help define objectives and measurable goals
- Rewrite programs into community-based descriptions
- Translate biblical terms into secular equivalents
- Provide feedback on tone for specific audiences

Sample Prompts to Use

- Review the following grant draft for clarity and professionalism. Suggest edits to make it concise and persuasive without changing the meaning.
- Summarize our feeding ministry for a donor newsletter.
- Describe discipleship using terms related to mentoring.
- Help draft a project overview for our youth summer camp.

Stewardship as Witness

Faithful stewardship is not only about fundraising. It is a testimony. When ministries manage resources well, communicate with clarity, and lead with excellence, they build trust. Stewardship becomes part of the Church's witness to both spiritual and civic communities.

Economic Empowerment as Kingdom Work

Churches can move from survival to transformation when they embrace economic empowerment. This includes:

- Teaching financial literacy
- Supporting local entrepreneurs
- Creating job-training or apprenticeship programs
- Investing in affordable housing or small business incubators

Grant funding can serve as a bridge to long-term change, both inside the church and throughout the neighborhood.

A Word of Caution

Stay rooted in your mission. Do not let financial opportunities pull your focus away from what God called you to do. Let clarity guide your communication, not compromise. Never share confidential financial or legal data with any AI tool. Use them for writing assistance only.

Always have experienced advisors review your proposals and financial plans before submission.

Final Encouragement

You already have the heart, vision, and calling. What you may need now is clarity of language and confidence in new systems. Let your message match the excellence of your ministry. Let your proposals carry the same passion as your preaching. And let your words open the doors your faith has already prepared.

Chapter 12 – Marketplace Ministry

WHEN WORK FUNDS THE WITNESS

> *"You yourselves know that these hands of mine have supplied my own needs and the needs of my companions."*
> — Acts 20:34 (NIV)

Ministry That Multiplies Through Work

Paul was a tentmaker, yet his hands did more than provide for himself. They financed the mission and lifted burdens from others. Today, many pastors carry the same call: working outside the church while shepherding God's people.

For the bi-vocational leader, this can feel like double duty. With the rise of artificial intelligence, the load can be transformed into double opportunity. AI is not only a time-saver. It is a revenue-generator. It empowers pastors to innovate in the marketplace, create new streams of income, and equip their people with skills that multiply Kingdom impact.

Marketplace work is not a distraction from ministry. It is part of the mission.

Why This Matters

1. **Provision Fuels Vision**
   Churches and ministries flourish when finances are strong.

Revenue generated through AI-supported businesses can fund outreach, staff, and community projects.

2. **Do More with Less**
AI helps pastors maximize their limited time, automating repetitive work and freeing capacity for prayer, study, and relationships.

3. **Empowering People**
AI is not only for leaders. It is also a tool for training members, developing their marketplace skills, and increasing their financial stability.

4. **A Witness of Innovation**
When pastors model ethical and creative use of AI in business, they demonstrate how technology can be redeemed for Kingdom purposes.

How AI Can Support Marketplace and Entrepreneurial Work

1. Freelancing and Creative Services

- Write blog posts, devotionals, and leadership articles.
- Manage content for businesses or ministries.
- Edit, proofread, and repurpose sermons into saleable content.

```
Prompt Example:
Draft a 750-word blog article on servant
leadership in business, optimized for
LinkedIn.
```

2. Digital Products and Online Courses

- Build e-learning experiences on faith, leadership, or financial stewardship.
- Convert sermons into workbooks or eBooks.
- Package devotional series for sale online.

> **Prompt Example:**
> Outline a 5-module online course for Christian entrepreneurs on managing time, faith, and finances.

3. Consulting and Coaching

- Use AI to draft proposals, contracts, and business strategies.
- Summarize industry trends to advise clients.
- Generate assessment tools and surveys.

> **Prompt Example:**
> Write a professional proposal for consulting services to help nonprofits improve fundraising and donor engagement.

4. E-Commerce and Side Businesses

- Generate product descriptions and marketing content.
- Plan launch campaigns and customer emails.
- Create branding, slogans, or content calendars.

> **Prompt Example:**
> Write 10 creative t-shirt slogans based on the theme of faith and perseverance.

5. Real Estate, Trades, and Services

- Create polished marketing materials for listings or small businesses.
- Automate client communications and follow-ups.
- Summarize legal or technical documents into plain language.

> **Prompt Example:**
> ```
> Create design mockups to turn this room into a
> training center with 3 desk and 9 chairs.
> [attach photo].
> ```

Cost Savings and People Development

AI reduces overhead by eliminating the need for multiple assistants, researchers, or designers. But beyond cutting costs, the true power lies in developing people.

Pastors can use AI to train volunteers and members, equipping them with skills in writing, design, marketing, and administration. These skills benefit the church while also giving individuals opportunities in the marketplace.

This creates a cycle of empowerment:

- The church saves money by building in-house capability.
- Individuals gain valuable, income-generating skills.
- Families grow in financial stability.
- A more prosperous congregation can better finance Kingdom work.

Here is a Kingdom truth: the Kingdom advances when God's people are equipped and resourced to give. By equipping members with AI-powered skills, pastors raise up a church body that is resourced, innovative, and ready to fund the mission.

> *Prompt Example:*
> *Create a simple training session outline to teach volunteers how to use AI for writing weekly newsletters.*

Benefits of Using AI for Revenue Generation

- **Scalability**: What once took hours can now be done in minutes, serving more clients with less effort.
- **Cost Savings**: AI reduces the need for large staff and outsourced tasks.
- **Speed to Market**: Move from idea to product faster, opening new income streams.
- **Focus on High-Value Work**: Delegate routine tasks to AI so pastors and leaders can focus on vision, people, and growth.
- **Community Empowerment**: Equip members to increase their own earning potential, strengthening the financial base of the church.

Reflection Questions

- What opportunities has God placed in my hands that AI could help multiply?
- How could I use AI not just for myself but to equip my people with marketable skills?
- What products, services, or insights am I already stewarding that could generate revenue through AI support?
- In what ways can my work outside the pulpit also become part of my witness?

Encouragement for the Bi-vocational Leader

You are not carrying two separate callings. You are carrying one multiplied mission. Your work and your worship are not divided. They are united under Christ.

AI is a tool that allows you to do more with less, generate provision for your ministry, and empower your people to rise financially. As they prosper, the Kingdom advances.

The bi-vocational leader is a pioneer. You are showing the Church how to shepherd faithfully while building creatively in the marketplace.

PART IV – BUILDING THE FUTURE CHURCH

Chapter 13 – Vision Builders: Church Planting & Innovation

EMBRACING NEW GROUND WITH FAITH AND STRATEGY

> *"See, I am doing a new thing! Now it springs up; do you not perceive it?"*
> — Isaiah 43:19 (NIV)

Starting Something New for the Kingdom

Church planting has always been an act of bold faith. Whether launching a new congregation in a living room, a storefront, or a school auditorium, church planters are visionaries. They sense the movement of God and respond by creating spiritual homes in places where none yet exist.

In this generation, the tools available to planters are changing. Alongside prayer, fasting, and leadership, we now have artificial intelligence—tools that can support innovation, streamline planning, and strengthen communication. Church planting still requires obedience and spiritual grit, but AI can assist in organizing, scaling, and sustaining the work.

This chapter explores how AI can serve pioneers and innovators without replacing the Spirit-led calling that undergirds every faithful step.

Why Innovation and Church Planting Go Hand in Hand

Innovation in the Church is not about trend-chasing. It is about discernment. Where is God moving? What new communities are forming? What spiritual hunger is rising?

Church planting meets these moments with structure and hope. It recognizes the spiritual need in a neighborhood, demographic, or digital space—and answers with the presence of God's people. Innovation ensures the methods remain relevant, even while the message remains eternal.

Planting churches today means navigating everything from launch teams and fundraising to social media, outreach strategy, and sermon planning. AI can help lighten the administrative and creative load so that leaders can focus on shepherding people.

Ways AI Can Support Church Planters

1. Community Research and Needs Assessment

AI can help gather and summarize demographic data, cultural trends, and community needs to shape a contextual launch plan.

> **Prompt Example:**
>
> ```
> Summarize community challenges and
> opportunities for a church plant in an urban
> neighborhood with high renter populations and
> a growing immigrant community.
> ```

2. Vision Casting and Communication

AI can help refine vision statements, core values, and mission language that resonates clearly and consistently.

> **Prompt Example:**
>
> ```
> Help write a compelling one-paragraph vision
> statement for a new church focused on
> reconciliation and racial unity.
> ```

3. Team Development and Training Materials

Assembling a launch team requires clear roles, onboarding processes, and leadership development. AI can assist in creating training materials and schedules.

> **Prompt Example:**
>
> ```
> Generate an outline for a 3-week launch team
> training series with Scripture, discussion,
> and tasks.
> ```

4. Budget Planning and Grant Writing

AI can assist with basic financial templates, sample budgets, or even the first draft of a funding proposal or grant application.

> **Prompt Example:**
>
> ```
> Create a sample one-year church plant budget
> for a congregation of 50 people meeting in a
> rented space.
> ```

5. Outreach Campaigns

From designing flyers to writing social media posts, AI can help plan and launch community engagement strategies tailored to your audience.

> **Prompt Example:**
>
> ```
> Write a 5-post social media campaign
> introducing a new church launching this fall
> in a college town.
> ```

6. Follow-up and Volunteer Coordination

AI can help draft follow-up emails for first-time guests, automate communication flows, and organize teams for setup, worship, kids' ministry, and more.

> **Prompt Example:**
> ```
> Write a follow-up email to someone who
> attended our interest meeting and expressed
> curiosity about getting involved.
> ```

Innovation That Honors the Mission

AI cannot plant a church. It cannot share a meal with a lonely neighbor, or preach with Spirit-filled conviction. But it can help you build systems and communication channels that make space for those very things to happen.

The true innovation is not in using tools. It is in how we respond to the needs of people and the call of God. When churches are planted with vision, humility, and strategic care, they become lifelines of hope and healing.

Building What Lasts

You are not just launching a service. You are planting a legacy. Each person you reach, each prayer you pray, and each seed you sow becomes part of a living work of God. Let the Spirit lead, let wisdom guide, and let the tools available serve you—not define you.

Innovation is not the goal. Obedience is. Faithfulness is. Transformation is. May every new church planted through your leadership be rooted in prayer, grounded in truth, and shaped by God's heart for His people.

Let your vision be bold. Let your planning be wise. And let your innovation serve the gospel, not the other way around.

Chapter 14 – Wisdom in a Wired World

SPIRITUAL DISCERNMENT & GUARDRAILS

> *"My sheep hear my voice, and I know them, and they follow me."*
> *— John 10:27 (ESV)*

Technology Is a Tool, Not a Shepherd

AI is not inherently good or evil. Like many tools, it reflects the purpose and posture of the person who uses it. While it offers new possibilities for ministry, it also invites us to ask deeper spiritual questions.

Just because a tool can do something does not mean it should. Just because AI can speak does not mean it speaks for God. In a world filled with voices, the sheep must learn to recognize the voice of the true Shepherd.

The Role of Discernment in Digital Spaces

Discernment is the spiritual ability to separate what is helpful from what is harmful, what is truth from what only appears true. In the age of AI, discernment becomes even more important.

Tools like ChatGPT can offer fast answers, inspiring words, or persuasive messages. But not everything that sounds good is grounded in God's Word. AI can echo biases, generate shallow advice, or reinforce unhealthy patterns if it is not guided carefully.

The role of the pastor, teacher, or leader is to test every message, examine every suggestion, and listen closely for what aligns with God's truth.

Questions to Guide Spiritual Discernment

When using AI tools in ministry, ask:

- Does this reflect the character of Christ?
- Is this faithful to Scripture?
- Does this point people toward truth, love, and hope?
- Would I share this with confidence in front of my congregation?
- Have I prayed over this message or material before presenting it?

These simple questions can act as guardrails, keeping your use of technology from drifting into distraction or deception.

Healthy Boundaries Around AI Use

Here are a few guardrails to help keep your engagement with AI spiritually healthy and mission-aligned:

- **Keep Human Oversight**
  Never rely fully on AI to speak for your church or ministry. Always review, refine, and approve content before sharing it. The reality is that AI can have you looking foolish.

- **Protect Sensitive Information**
  Do not upload personal, financial, or pastoral care details into AI systems. Use general descriptions and keep confidentiality sacred.

- **Use It to Support, Not Replace**
  Let AI assist with writing, organizing, or brainstorming, but keep your heart, voice, and prayer at the center of all you produce.

- **Teach Your Team and Community**
  Help your staff and members understand how AI works, where it can serve, and where it must be limited. Equip others to think critically and faithfully.

- **Stay Rooted in the Word and in Prayer**
  No matter how advanced technology becomes; it will never

replace the presence of the Holy Spirit or the authority of Scripture. Keep your devotional life strong so that your discernment remains clear.

Prompt Examples for Reflection and Wisdom

```
Help write a devotional thought on trusting
God during seasons of uncertainty, rooted in
Psalm 23.
```

```
Create a short list of reflection questions
for a leadership team navigating big
decisions.
```

```
Suggest ways to present the concept of digital
wisdom to a youth group.
```

These prompts can help you explore topics, but only prayer, Scripture, and the Spirit will give you the wisdom to teach them well.

Listen Well

As you continue exploring AI in ministry, do not move so quickly that you lose the habit of listening. God's voice still speaks in the quiet place, in the prayerful pause, and through His Word.

I have repeated this thematically across the book as often as I could. It could not be restated enough: AI may offer answers, but only God offers truth. AI may save time, but only God redeems it. AI may generate content, but only God transforms hearts.

Let every tool you use in ministry be submitted to the One who called you. Do not be afraid of new technologies, but do not be distracted by them either. Let wisdom lead the way.

Chapter 15 – Walking Justly in the Age of AI

ETHICS, RESPONSIBILITY, AND MINISTRY INTEGRITY

> *"He has shown you, O man, what is good; and what does the Lord require of you but to do justly, to love mercy, and to walk humbly with your God?"*
> — Micah 6:8 (NKJV)

Technology Reveals the Heart

In addition to being a tool for productivity and convenience, AI is also a mirror. How we use technology reveals what we value. Our decisions about AI can either reflect justice, compassion, and humility—or expose our blind spots and biases.

As pastors, leaders, and believers, we are called to walk justly in every area of life. This includes how we use digital tools. Ethics in ministry is not a separate concern from innovation. It is the foundation that guides everything we do, especially when the world is watching.

AI and the Moral Landscape

AI tools can be used to care for people, teach truth, and extend the reach of the gospel. But they can also be misused in ways that harm, deceive, or exclude.

Some of the ethical concerns surrounding AI include:

- Plagiarism and uncredited content
- Deepfakes or altered media used to manipulate perception
- Misinformation and the rapid spread of false narratives

- Automation that replaces human connection or jobs without care
- Data privacy violations
- Reinforcement of social bias or injustice in generated content

These are not just technical issues. They are spiritual and moral concerns. They affect how we lead, how we serve, and how we reflect the character of Christ.

Practicing Justice with AI Tools

Justice in ministry means more than fairness. It means using your influence, resources, and decisions to reflect God's righteousness. Here are ways to practice justice when using AI:

1. **Be Honest**

    If AI helps you write or create, acknowledge it. Do not pass off automated content as purely your own without review or prayer. Give credit where it is due and remain transparent about your creative process.

2. **Be Inclusive**

    Check the language and tone of your content. Make sure it reflects a wide range of people and avoids cultural bias. AI models are trained on existing data, which may carry assumptions you need to challenge.

3. **Be Cautious with Representation**

    Avoid generating images or videos that misrepresent real people. Do not use AI-generated likenesses for manipulation or emotional influence. Be clear about what is real and what is simulated.

4. **Be a Guardian of Privacy**
    As mention in an earlier chapter: never input private details, counseling notes, or confidential information into AI platforms. Protect the dignity and trust of the people you serve.

5. **Be Accountable**

   Invite others into your creative and strategic process. Let elders, peers, or ministry partners help you evaluate how AI is used in your church or organization.

Ethics as an Act of Worship

Ethical leadership is not just about protecting your reputation. It is about honoring God in everything you do. When you choose the right path even when no one is watching, you bear witness to the holiness of your calling.

Your use of technology is part of your testimony. Let it reflect humility, compassion, and wisdom. Let it build bridges instead of barriers. Let it honor the image of God in every person.

Reflection Prompts for Ethical Discernment

```
Help me write a code of ethics for AI use in
church media and teaching.
```

```
Suggest five questions a church staff should
ask before adopting AI tools.
```

```
Create a devotional thought on truthfulness
and transparency in leadership.
```

```
Suggest a policy outline for how our ministry
can responsibly use AI.
```

These prompts can help shape internal conversations and external policies rooted in biblical values.

Walking Justly Is a Witness

You do not need to fear AI, but you do need to walk wisely. Let your decisions reflect the heart of Christ. Let your actions protect the vulnerable. Let your creativity be anchored in truth. Justice is not just a social issue. It is a gospel issue.

In every age, the Church has faced new ethical challenges. This one is no different. The opportunity is before you to lead with courage, clarity, and conviction.

Let your walk in the digital world be marked by the same mercy and justice that shaped the steps of Jesus.

Chapter 16 – Watchmen on the Wall

SECURITY, PRIVACY, AND DIGITAL TRUST

> *"Whoever walks in integrity walks securely, but whoever takes crooked paths will be found out."*
> —Proverbs 10:9 (NIV)

The Weight of Trust

Churches are holy spaces. Beyond the walls of worship, they are places where trust is given and lives are shared. Members bring their deepest hurts, hopes, and needs to their leaders. As ministry enters digital spaces—and especially when incorporating artificial intelligence—security and privacy must be treated as ministry essentials.

Whether prayer requests, counseling conversations, or giving records, every bit of data reflects a person and their spiritual journey. Mishandling this information isn't just a technical failure—it's a breach of pastoral care.

Why Security and Privacy Matter More Than Ever

Churches today are not immune to the challenges facing digital organizations. Hackers, data leaks, and phishing scams have targeted nonprofits and churches alike. AI tools, while powerful, often process and store data in the cloud. If leaders aren't intentional, sensitive church information can be at risk.

Many church leaders may not realize that AI tools come with terms of service and privacy policies that should be reviewed. Settings within tools like ChatGPT may include options to share data with developers for model improvement. While this may sound harmless, it can put private church information at risk.

Always verify and adjust the settings of any AI tool your church uses. Turn off optional data sharing. If you're not sure, consult someone with tech expertise. You wouldn't leave the church doors unlocked at night—don't leave digital doors open either.

What Not to Enter into AI Tools

Even when privacy settings are managed well, wisdom calls for restraint. Don't enter:

- Real names or identifying details of counseling or pastoral care situations
- Giving records, financial pledges, or donation information
- Medical, legal, or marital updates shared in confidence
- Passwords or login information

Instead, generalize. Reword requests and scenarios to protect identity and without violating someone's trust.

How Security Breaches Happen

Breaches occur in many ways:

- Weak or shared passwords
- Open accounts that are no longer monitored
- Cloud-based tools storing unprotected data
- Inputting sensitive content into AI tools without understanding where it's stored

Breaches aren't always dramatic—they can be slow leaks of trust. One mishandled message, one forwarded file, or one chatbot conversation shared too publicly can sow seeds of distrust that take years to uproot.

Understanding Government Regulations

Government laws are increasingly holding organizations accountable for digital responsibility. Some of the regulations churches should be aware of include:

- GDPR (General Data Protection Regulation): Applies primarily in the European Union but influences global practices. It emphasizes user consent, the right to be forgotten, and secure data handling.

- HIPAA (Health Insurance Portability and Accountability Act): While most churches aren't covered entities, if any form of health-related information is stored or shared (such as in counseling), aspects of HIPAA may apply.

- CCPA (California Consumer Privacy Act): Provides guidelines on how data of California residents must be handled. Similar laws are spreading to other states.

Even if not legally required, it's wise to align with the spirit of these protections. They reflect values churches should already embrace: transparency, dignity, and respect.

Best Practices for Church AI Use

To keep your church safe and your people's trust intact, consider these safeguards:

- Review and adjust AI tool settings. Turn off permissions to share data with model trainers or developers.

- Limit access. Only allow trusted staff to access AI platforms, and never share passwords casually.

- Create a digital use policy. Clearly state what kinds of data can and cannot be used with AI tools.

- Audit regularly. Review who has access, what information has been shared, and whether settings have changed.

- Train your team. Ensure that staff and volunteers understand the boundaries and responsibilities of digital ministry.

Proceed with Caution

Security and privacy are not just technical concerns—they are spiritual ones. When you guard what's sacred, you mirror the heart of the Good Shepherd who watches over His flock.

The digital tools in our hands can build trust or break it. Used wisely, they become vehicles for grace. Used carelessly, they can compromise the very message we're called to share.

Lead with integrity. Set the example. And remember: true safety is found not only in settings and passwords, but in the loving care we show toward every story, every soul, and every sacred piece of information entrusted to us.

BRIDGING THE PAST AND FUTURE

BRIDGING THE PAST AND FUTURE

Chapter 17 – The Cloud of Witnesses: Preserving Our Stories

TESTIMONIES AND ELDERS IN A DIGITAL AGE

> *"Therefore, since we are surrounded by so great a cloud of witnesses, let us lay aside every weight... and let us run with endurance the race that is set before us."*
> — Hebrews 12:1 (NKJV)

Stories Are Sacred

In every generation, the Church has been carried forward not only by doctrine and teaching, but by story. Testimonies are the living echoes of God's faithfulness. They show how grace has touched lives in real, personal, and unforgettable ways.

Elders and long-time members hold these stories like treasures. Their voices, seasoned by experience and faith, anchor the body of Christ in continuity. These testimonies are more than memories. They are lessons, encouragements, and legacies that must be preserved.

The Risk of Silence

If we do not record these stories, they fade. If we do not ask the questions, they go unanswered. If we do not preserve the wisdom of our elders, we risk losing a part of our identity as a spiritual family.

The Church is not only what is happening now. It is also what has happened before and what will happen next. Testimonies create connection between generations. They build bridges between struggles and victories, between tradition and innovation, between past and future.

Why Preservation Matters Now

In this digital age, we have more tools than ever to capture, store, and share stories. But unless we are intentional, those tools can remain unused. The urgency is real. Elders age. Moments pass. Memories fade.

Preserving stories now means creating a legacy that your children's children can look back on. It means giving people in your church the opportunity to hear how God moved before they arrived. It means honoring the ones who paved the way and showing the next generation that they are part of something much bigger.

Ways to Preserve Testimonies and Elders' Wisdom

** Obtain the necessary permission and consent from the person, guardian, etc. **

1. **Recorded Interviews**

    Set up short interviews with older members or those with significant testimony. Ask about their spiritual journey, the church's history, or how they have seen God at work. Record these sessions for future use in worship, training, or archives.

2. **Written Stories**

    Invite members to write out their testimonies or submit typed reflections. These can be gathered into newsletters, devotionals, or anniversary books.

3. **Video Clips for Services**

    Share 1 to 2-minute stories during worship or online as part of encouragement segments. These create connection and stir faith.

4. **Testimony Nights or Digital Testimony Walls**

    Hold evenings where stories are shared live or set up a section on your church's website or app for testimonies.

5. **Create a Church Memory Archive**

Use cloud storage to preserve audio, video, and written materials for future generations. Label and organize them clearly for access.

How AI Can Help Preserve Stories

AI can be a supportive tool in collecting and presenting these stories. Here are ways it can assist:

- Generate interview questions for elders, covering life, faith, and wisdom
- Organize written testimonies into polished paragraphs or devotional formats
- Summarize long audio interviews into key quotes or highlights
- Help write introductions or transitions for video story segments
- Draft invitations or announcements to encourage people to share their stories
- Turn scattered notes or old recordings into tribute articles or memory books

ChatGPT does not replace the storyteller. It supports the work of honoring and recording the story.

Prompt Examples to Use

```
Create a list of 10 questions to ask elders in
a testimony interview.
```

```
Rewrite this handwritten testimony into a
devotional for our newsletter.
```

```
Summarize this 20-minute testimony into a
paragraph with key takeaways.
```

```
Draft a thank-you note to a member who shared
their life story publicly.
```

Write an intro script for a video sharing testimony from church pioneers.

Let the Witnesses Speak

Your church's story is not just found in sermons or bulletins. It is found in the quiet faith of the widower who never stopped serving, in the tears of the mother who prayed her children back to the Lord, and in the joy of the teenager whose life was changed at camp.

Preserving these stories is an act of worship and stewardship. You are not only collecting memories. You are recording miracles.

Let your church be one that listens well, remembers deeply, and honors those who have walked before. Do not let the cloud of witnesses go unheard. Give them voice. Give them space. Let their testimonies fuel the faith of the next generation.

Chapter 18 – Beyond the Horizon

PROCLAIMING THE WORD IN A DIGITALLY DISCERNING AGE

> *"Preach the word; be prepared in season and out of season; correct, rebuke and encourage—with great patience and careful instruction."*
> — 2 Timothy 4:2 (NIV)

The Church at the Edge of Innovation

Throughout history, the Church has stood as both a witness to change and a voice of truth within it. Whether it was the printing press, the radio, or the livestream, the Church has continually found ways to proclaim the unchanging gospel through ever-changing tools.

Today, artificial intelligence is one of the most significant technological shifts of our time. It can support the work of ministry in powerful ways. To be faithful in this moment, we must not only understand the technology but also learn how to guide its.

How AI Is Already Supporting Ministry

AI is not about replacing spiritual leadership; it is about enhancing the ways we serve. Some churches are already using AI to:

- Organize sermon research
- Translate services into multiple languages for multicultural congregations
- Create resources for Bible study and discipleship

- Help community members write resumes, access services, or learn digital skills
- Streamline communication, outreach, and administration

These examples are not science fiction. They are happening now. The opportunity before us is to steward these tools faithfully and to expand access, compassion, and clarity in our ministries.

The Church as a Moral and Missional Voice

Technology is never neutral. It reflects human intention. That is why the Church must step into the broader conversation about AI—not just to protect its own interests but to help shape public life with values that reflect the kingdom of God.

Churches and Christian leaders can:

- Influence how AI is used in schools, hospitals, and justice systems
- Advocate for ethical standards that protect the vulnerable
- Equip believers to approach technology with discernment and faith
- Raise up leaders who understand both Scripture and systems, theology and technology

AI affects not just how we communicate, but how we make decisions, distribute resources, and relate to one another. The Church's presence in these spaces can bring light to areas that are too often driven only by profit or power.

Preaching with Courage and Clarity

In this moment of technological change, the Church's voice matters more than ever. Preachers are not just speakers—they are interpreters of culture, shepherds of souls, and ambassadors of hope. AI can assist in study, structure, and preparation, but it cannot deliver the Word with the authority and anointing of someone called and commissioned by God.

Preaching in this season means addressing real questions people are asking about ethics, automation, purpose, and identity. It means lifting up a vision of God's kingdom that brings transformation—even in digital spaces.

A Theology of Hope

The book of Revelation promises a new heaven and a new earth—not powered by machines, but by the presence of God. That future is not something we must manufacture. It is something we wait for, prepare for, and witness to.

The Church does not place its hope in technology. Our hope is in the One who makes all things new. Yet in the meantime, we are called to be good stewards of the tools in our hands. AI is one such tool.

Final Words of Encouragement: Lead with Light

The future of the Church is not mechanical. It is missional.

Let every tool in your hand be directed by the Spirit in your heart. Let every innovation be tempered by wisdom and compassion. Let your preaching continue to stir hearts, challenge systems, and reveal Jesus in every age; including this one.

AI is not the future of the Church. Jesus is.

So go forward in faith. Proclaim the Word. Shape the future. And trust that the God who began a good work in us will carry it on, no matter what tools the world may bring.

Chapter 19 – God's Gift: Man, Not Machine

WHEN NOT TO USE AI

> *"What is mankind that you are mindful of them, human beings that you care for them? You made them a little lower than the angels and crowned them with glory and honor."*
>
> — Psalm 8:4–5 (NIV)

Why AI Has Its Limits in the Sanctuary

Artificial intelligence is powerful when it comes to efficiency, research, and productivity. However, in the sacred space of worship, its limits become very clear. Worship and preaching are not only about words on a page. They are about spirit, presence, and the relational bond between God's servant and God's people. Replacing these divine elements with mechanical outputs risks creating a hollow substitute that lacks the breath of life.

Case Study 1: Violet Crown City Church — Austin, Texas

In September 2023, Violet Crown City Church, a United Methodist congregation in Austin, Texas, conducted an unusual experiment. The entire worship service was designed with the help of ChatGPT. The pastor prompted the AI to create prayers, write a sermon, suggest songs, and structure the order of worship. What resulted was a full service that reflected the style of mainline Protestant liturgy and sounded polished on paper[12].

Although the experiment drew attention, the response was telling. Some members appreciated the novelty, but many felt something essential was missing. The prayers lacked warmth and depth. The sermon, while grammatically correct, seemed shallow and without the urgency that comes when a preacher proclaims a message from time in prayer and study. The pastor himself admitted that the service felt hollow, noting that AI is "not spirit-empowered"[3].

Afterward, the church decided not to repeat the experiment. The lesson learned was that while AI could generate liturgical content, it could not breathe life into it. The people of God discerned the difference between well-structured words and Spirit-filled preaching.

Case Study 2: St. Paul's Church, Fürth, Germany — Kirchentag 2023

At the Evangelical Kirchentag in Fürth, Germany, in June 2023, a highly publicized experiment took place. A team of theologians and computer scientists used ChatGPT to generate nearly an entire worship service. Prayers, blessings, a sermon, and liturgical elements were written by AI and delivered by digital avatars projected on a large screen. More than 300 people attended to see how artificial intelligence might function in a sacred setting[4,5,6].

---

[1] Patterson, J. (2023, November 21). *Artificial intelligence and church.* UM News. Retrieved from UM News website Evangelical Focus+8Popular Science+8World Religion News+8

[2] Valencia, K. (2023, September 19). *A Texas church service devised by ChatGPT as AI takes over worship, prayers and sermon.* Premier Christian News. Retrieved from Premier Christian News website

[3] Chamberlain, D. (2023, October 16). *Following AI-generated worship service, Texas pastor reflects on 'learning opportunity,' says AI is not 'spirit-empowered.'* ChurchLeaders. Retrieved from ChurchLeaders website

[4] Evangelical Focus. (2023, June 13). *First artificial intelligence-led worship service tested in Germany.* Evangelical Focus. Retrieved from Evangelical Focus website

[5] Grieshaber, K. (2023, June 10). *Can a chatbot preach a good sermon? Hundreds attend church service generated by ChatGPT to find out.* AP News. Retrieved from AP News website

[6] Ars Technica. (2023, June 12). *ChatGPT takes the pulpit: AI leads experimental church service in Germany.* Retrieved from arsTechnica.com

The reactions were deeply mixed. Some attendees admitted curiosity and even admiration for the technology. Yet many found the experience unsettling. The avatars appeared polished but lacked any sense of empathy. Congregants noted that the AI-generated sermon was repetitive and emotionless. One attendee described it as "a shallow shell of worship." Others felt disturbed that an avatar was attempting to lead them in the Lord's Prayer. Several participants refrained from speaking, feeling that such a sacred moment should not be led by a machine[7].

The experiment made global headlines. While it demonstrated what technology could do, it also revealed what technology should not do. Pastoral ministry is not simply about transmitting information. It is about presence, empathy, discernment, and the work of the Holy Spirit.

Pastoral Reflection and Church Response

These case studies reveal a vital truth. The church is called to encounter. AI can produce outlines, scripts, and avatars, but it cannot embody compassion. It cannot walk with a grieving family. It cannot pray with the conviction of a heart moved by the Spirit. It cannot preach with tears, laughter, or the holy authority that comes from divine calling.

The conclusion is simple. Technology is a tool. Pastors and leaders are entrusted by God to proclaim His Word with their whole being.

A Human-Centered Conclusion

People of God, YOU HAVE NO COMPETITION.

God has chosen humanity, not machines, to carry His message. Scripture tells us that He crowned us with glory and honor (Psalm 8:5). That glory is not transferable to algorithms. The Spirit speaks through flesh and blood.

Man is God's greatest gift. While AI may support the work of the church in administrative and educational ways, it must never replace the irreplaceable gift of the pastor's own voice.

---

[7] Premier Christian News (Texas church service by ChatGPT)

Pastors and leaders, God has placed within you a sound that only you can deliver. Your voice carries the authority of your calling, the weight of your testimony, and the breath of the Spirit. **Use your voice.** The world may offer substitutes, but heaven has entrusted you with a holy sound that cannot be replicated.

Reflection Questions

1. How do you discern where AI is helpful versus where it undermines spiritual authenticity?
2. In what ways can your voice minister that no machine ever could?
3. What steps will you take to guard the holy space of worship from artificial replacements?

The Principle

Be led by the Spirit in *all* things.

If you use AI, you can train and direct it. It can handle administrative tasks, research, and even help spark creative ideas, but you release the sound. Your sound carries divine authority. AI may serve your mission, but it can never embody your anointing.

☞ *"God anoints people, not algorithms."*

The anointing flows through your sound, not through code.

As you steward your voice, consider partnering with myself or someone else who can help you train AI to reflect your unique style. That way, it becomes a tool shaped by your message rather than a substitute for it.

Thank You

Thank you for taking the time to read *Church, It's Time to Try AI: Because the Marthas Want to Take a Seat Too*. I pray that these pages have sparked not only new ideas but also a deeper sense of freedom— freedom to lead with revelation without carrying on unnecessary weight of ministry.

AI is not the answer to everything, but it is helpful. Your calling, your heart, and the Spirit of God are what bring life to your work. My hope is that this book has simply given you a head start in imagining what's possible when faith and technology walk together.

May the introductory information gathered here provide you with practical next steps to equip and develop yourself and your people. The tools are here. And with God's help, your hands will be strengthened for the work ahead.

With gratitude,
Dr. D. MeShayle Lester

Please Leave a Review

Thank you for walking through these chapters with me.

If this book has been a blessing to you, I would be grateful if you would leave a book review on Amazon.

Your feedback not only encourages me as the editor, but also helps other pastors, church leaders and Kingdom entrepreneurs discover this resource for their own journey.

Together, we can continue to equip the Church for faithful innovation.

About the Author

Dr. D. MeShayle Lester has faithfully served pastors and churches for more than thirty years, offering trusted support in roles such as executive assistant, ministry administrator, and strategic advisor. She brings together deep ministry experience and technical insight to help overwhelmed leaders find clarity, build systems, and lead with peace.

As a licensed and ordained minister, Dr. Lester understands the weight of pastoral leadership. Her ministry philosophy is rooted in the example of Jesus Christ, *the ultimate servant leader,* who modeled rest, delegation, and focused ministry over exhaustion and burnout. Her work centers on sustainable leadership, healthy practices, and Spirit-led innovation.

Dr. Lester is also a certified AI consultant who is passionate about equipping churches to embrace technology in ethical and practical ways. Through her various organizations and community projects, she helps leaders use AI not as a replacement for ministry, but as a trusted tool for communication, creativity, and transformation.

www.ingramcontent.com/pod-product-compliance
Lightning Source LLC
LaVergne TN
LVHW051600080426
835510LV00020B/3066